はしがき

スタジアムを汚す鳥の糞、また、果樹園を荒らすカラス等を撃退する方法には、強力な磁気や磁石によるものは、ペースメーカーに与える影響がありますが、このドローンには、紫外線を遮断する装備がなされている。これにより、センサーで自動的にドローンが離着し、システムのテリトリーを巡回すると、カラス等は、紫外線に視界を遮断され、飛び去る。このドローンには、紫外線遮断装備の他、後頁に記載の撃退装備がなされている。

Preface

It is the method of repulsing the crow which damages the excrement of the bird which soils a stadium, or an orchard.
Some which are depended on a powerful magnet have the influence which it has on the pacemaker of a human body.
The equipment which intercepts an ultraviolet ray is made by this drone.
If a drone makes take-off and landing at a sensor automatically by this and the territory of a system is patrolled, a crow etc. will have a field of view intercepted by the ultraviolet ray, and will fly away.
Repulse equipment of a publication is made by the back page besides ultraviolet ray interception equipment at this drone.

目　次

⑴ スタジアムのカラス撃退システム

① スタジアムのシステム概念図

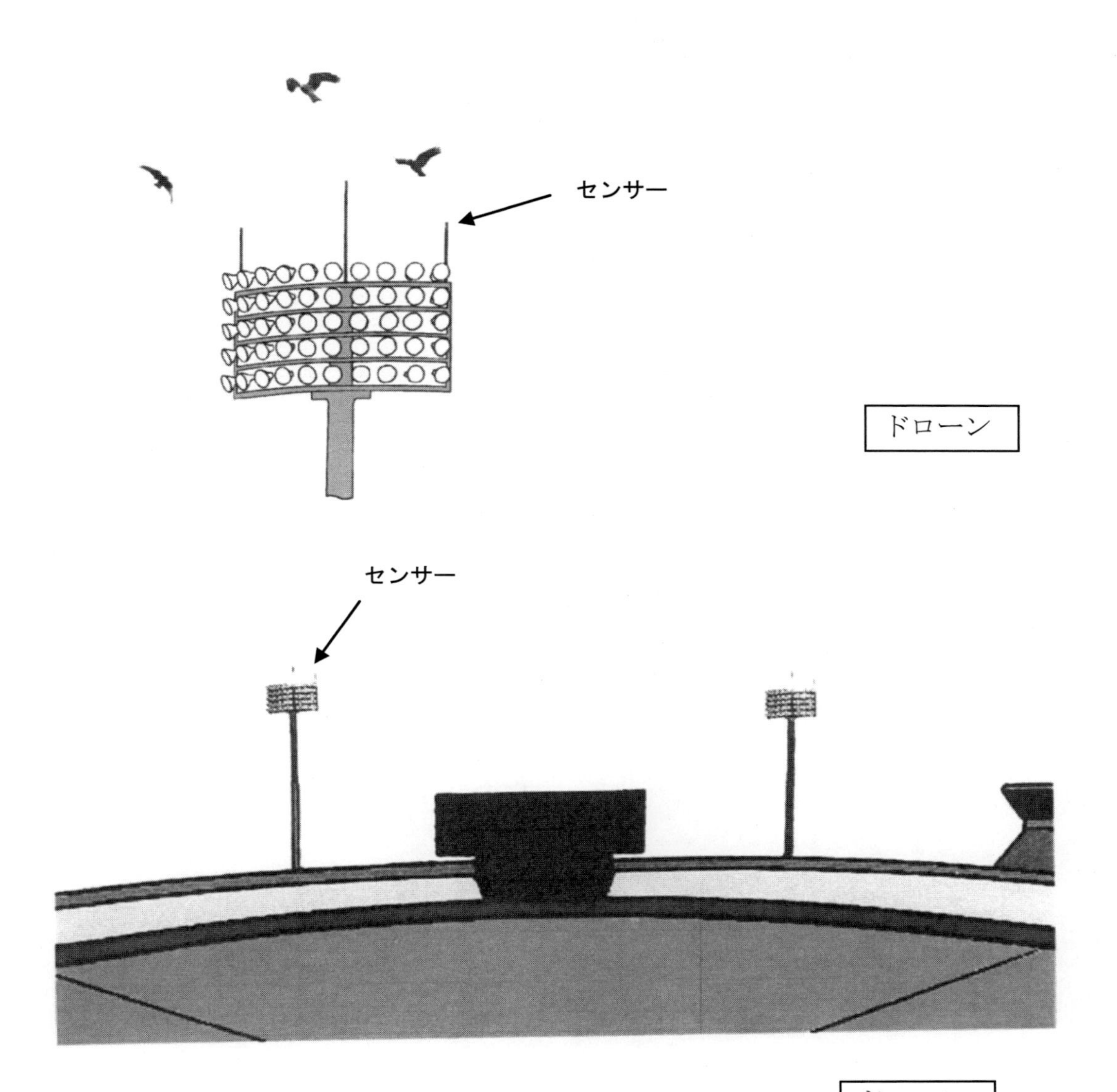

スタジアム内にカラスが飛来すると、センサーにより感知され、指令を受けたドローンが離着陸場から自動的に離着し、テリトリーを巡回するプログラム。また、手動でも可能。
ドローンには紫外線を遮断する装備がなされ、カラスの視界を遮断し、カラスを追い払う。
※宣伝・説明・明細書表現などへの引用を禁止。

② スタジアムの照明

スタジアムの照明塔にはカラスが集まり、糞公害が見られる。この多量の糞が風雨により、飛散し、感染症などの悪影響の原因でもある。
照明塔の糞の清掃は、容易に出来ない為、この問題を回避したのが、このシステムである。

※宣伝・説明・明細書表現などへの引用を禁止。

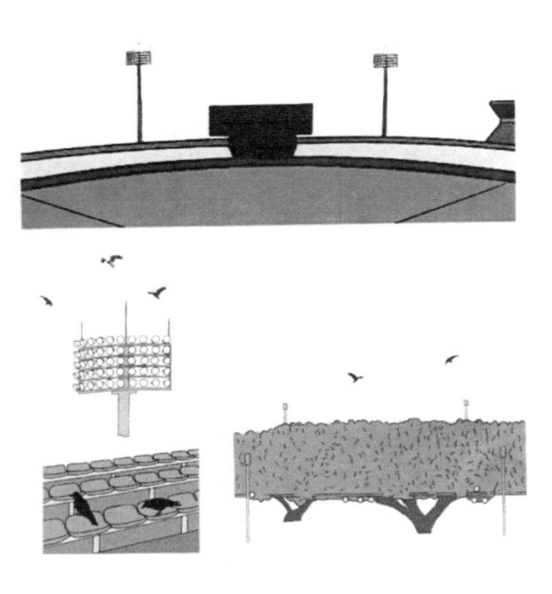

③　スタジアムのベンチ

ベンチの糞公害の清掃は容易でない。特にカラスの場合は、鳩などと異なり、執念深いから清掃の繰り返しである。
この問題を解決するのは、カラスをスタジアム内に飛来させないことであり、この目的を達成させるにはドローンを巡回させてカラスを追い払うシステムのプログラムである。

※宣伝・説明・明細書表現などへの引用を禁止。

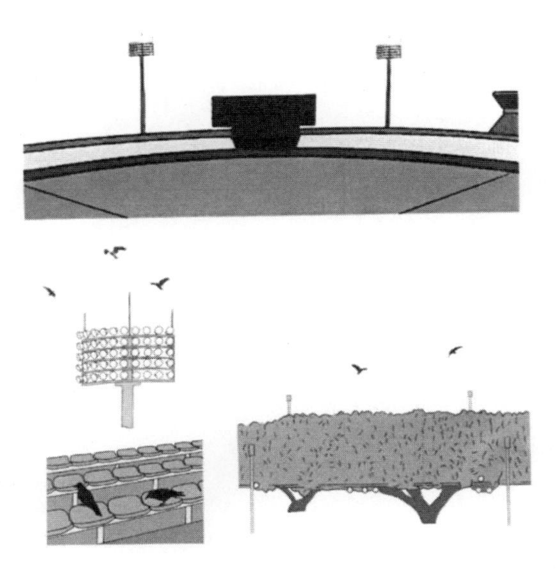

⑵　果樹園のカラス撃退システム

①　果樹園のシステム概念図

②　果樹園のドローン離着陸場

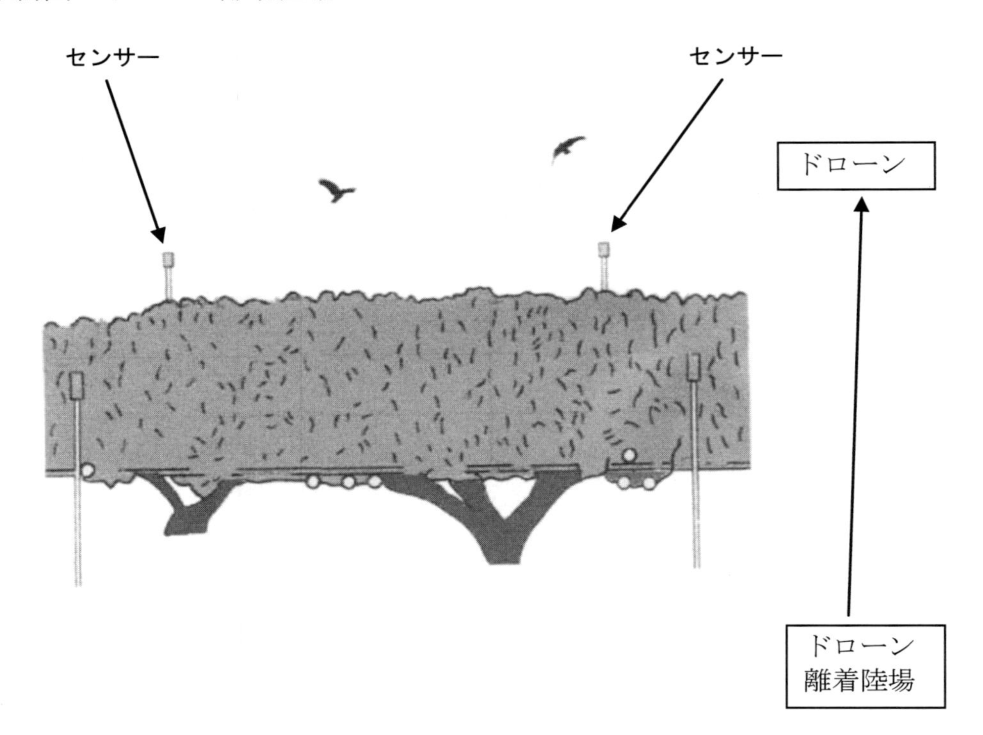

果樹園内にカラスが飛来すると、センサーにより感知され、指令を受けたドローンが離着陸場から自動的に離着し、テリトリーを巡回するプログラム。また、手動でも可能。
ドローンには紫外線を遮断する装備がなされ、カラス（鳥類）の視界を遮断し、カラスを追い払う。
果物収穫作業中にカラスが飛来し、ドローンがペースメーカーの作業者に接近しても、強力な磁気、磁石を装備してないから作業者への影響はない。

※宣伝・説明・明細書表現などへの引用を禁止。

(1)

Ultraviolet ray interception equipment drone crow repulse

①

The crow repulse system of a stadium

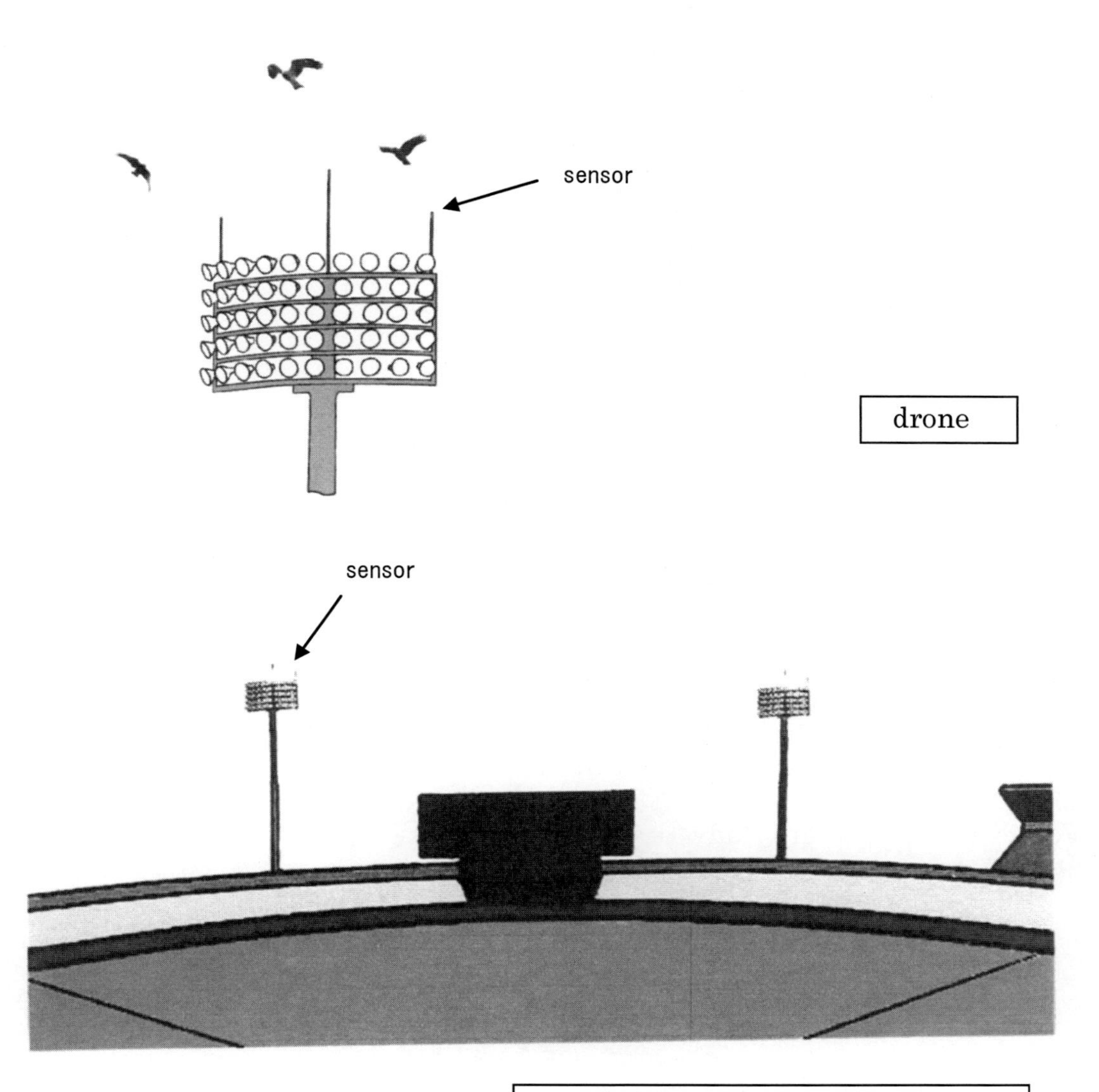

The program which it will be perceived by the sensor, and the drone which received instructions will fly automatically from a taking-off-and-landing place, and will patrol a territory if a crow comes flying in a stadium.

Also manually possible.

The equipment which intercepts an ultraviolet ray is made by the drone, the field of view of a crow is intercepted, and a crow is repulsed.

Forbid the quotation to advertisement, explanation, specification expression, etc.

②

Lighting of a stadium

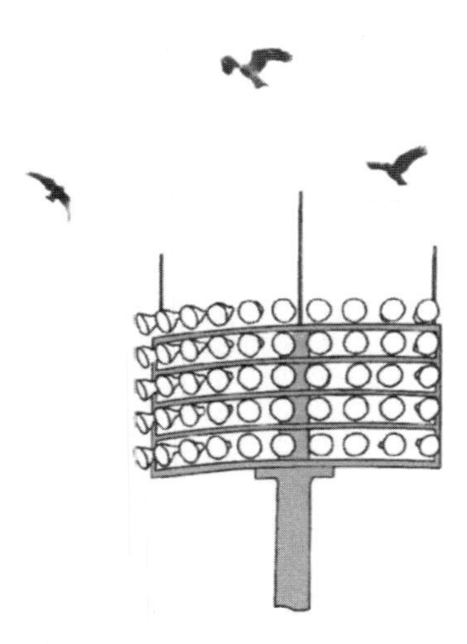

the lighting of a stadium -- a crow flocks in a tower and excrement pollution is seen

These excrement of a lot of disperses by the rainstorm, and also caused bad influences, such as infection.

lighting -- since cleaning of the excrement of a tower was not able to be performed easily, this system avoided this problem

Forbid the quotation to advertisement, explanation, specification expression, etc.

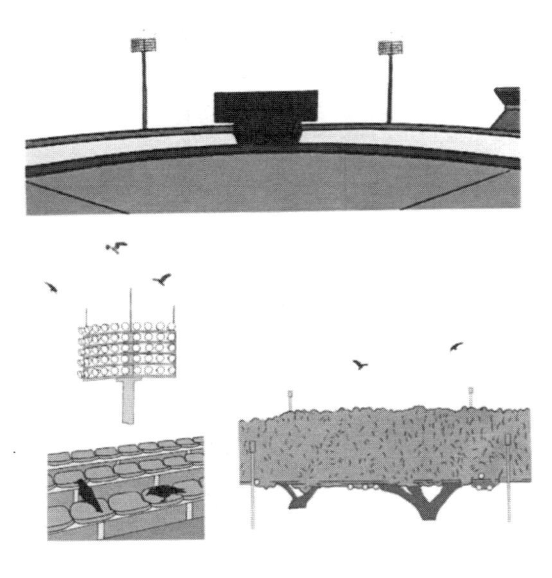

③

The bench of a stadium

Cleaning of the excrement pollution of a bench is not easy.
Since it is [in the case of a crow] especially persistent unlike a pigeon etc., it is a repetition of cleaning.
The program of a system which makes it patrol a drone to be not making a crow come flying in a stadium, and to make this purpose attain, and makes a crow repulse solves this problem.

Forbid the quotation to advertisement, explanation, specification expression, etc.

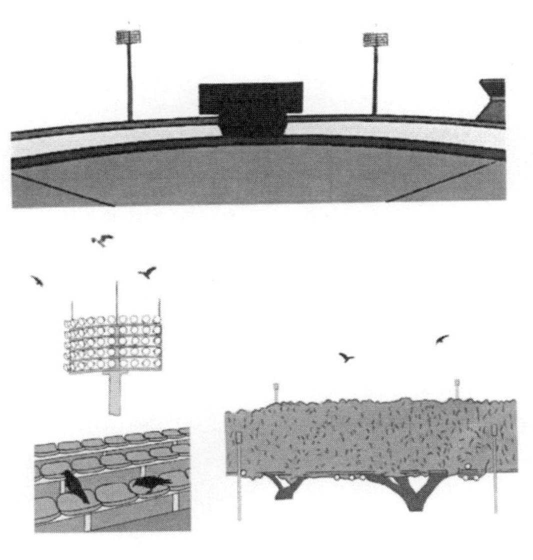

(2)

The crow repulse system of an orchard

①

The system concept figure of an orchard

②

The drone taking-off-and-landing place of an orchard

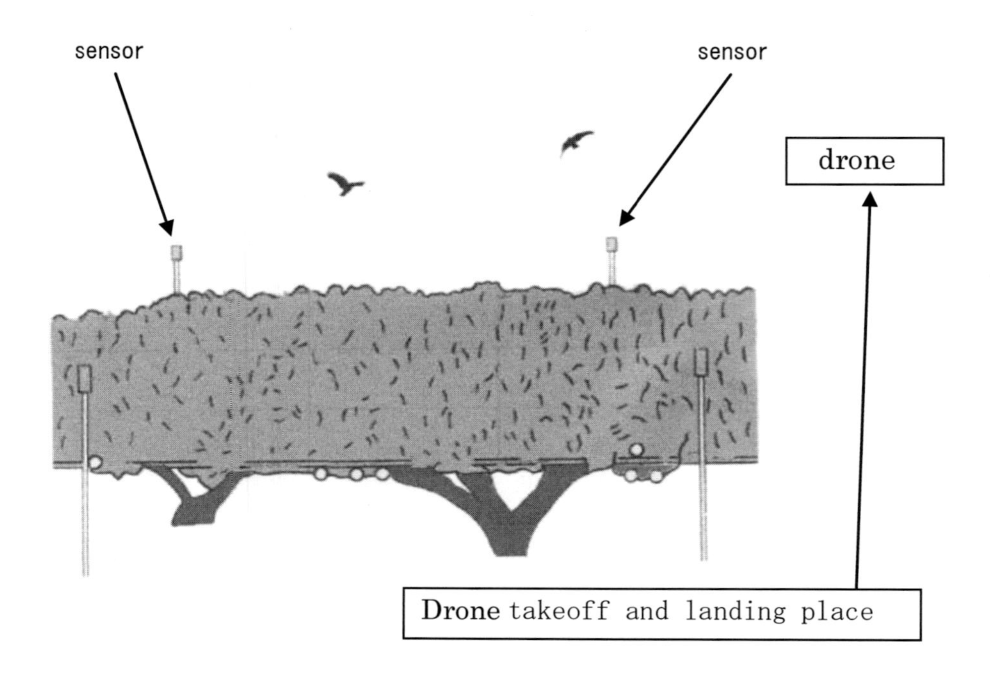

The program which it will be perceived by the sensor, and the drone which received instructions will fly automatically from a taking-off-and-landing place, and will patrol a territory if a crow comes flying in an orchard.
Moreover, also manually possible.
The equipment which intercepts an ultraviolet ray is made by the drone, the field of view of a crow (birds) is intercepted, and a crow is repulsed.
Since powerful magnetism and the magnet are not equipped even if a crow comes flying during fruit harvest work and a drone approaches a pacemaker's worker, there is no influence on a worker.

Forbid the quotation to advertisement, explanation, specification expression, etc.

⑶　公報解説

実用新案登録第 3199308 号

考案の名称；鳥獣害対策用の小型無人航空機

実用新案権者；樋口　節美

【要約】

【課題】畑等を荒らす鳥獣に対し、畑等の上空における飛行による動きや飛翔音或いは光等の威嚇手段を用いて威嚇し、撃退できる鳥獣害対策用の小型無人航空機を提供する。

【解決手段】使用時における水平方向へ十字状の腕部が延設された軽量かつ剛性を有する樹脂の中空筐体の中に、動力手段であるモーターと、動力源であるリチウムイオンポリマー二次電池と、機体の動作を制御する遠隔操作手段及び自動制御手段であるプロセッサ、通信装置、ＧＰＳ、磁力計、ジャイロスコープ、加速度計、超音波高度センサー、気圧計等が格納され、下面側に取着された２本の脚部１２と、上面側の先端側の各々に回転自在に設けられ、モーターからの動力を得て機体を飛行可能とする４つの回転翼１１とを有する本体部１０と、ホログラム加工のカッティングシートを本体部１０表面に貼り付けてなる反射面１３０とを備える。

【選択図】図２

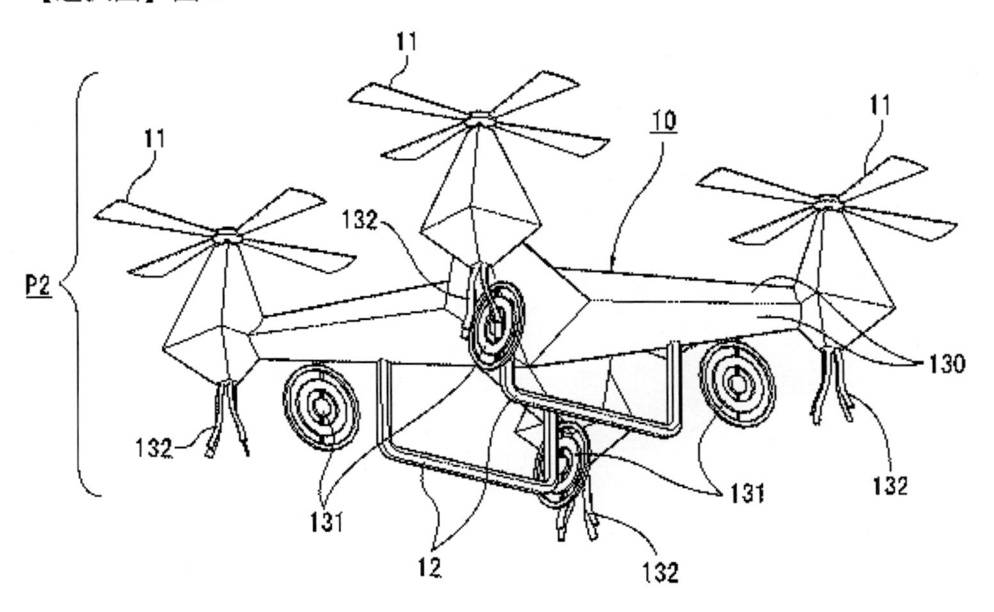

【実用新案登録請求の範囲】

【請求項１】

動力手段と、同動力手段を動かす動力源と、同動力手段から動力を得る推進手段と、機体の動作を制御する遠隔操作手段あるいは自動制御手段とを有する本体部と、

前記本体部に設けられた、鳥獣を威嚇するための威嚇手段とを備える

鳥獣害対策用の小型無人航空機。

【請求項２】

前記推進手段が回転翼である

請求項１に記載の鳥獣害対策用の小型無人航空機。

【請求項３】

前記本体部の自動制御手段が、自動離着陸機能と、高度設定機能と、飛行させる区域内の境界点座標値を設定することで飛行区域の位置情報を入手可能なＧＰＳ受信機能と有するものである

請求項１または請求項２に記載の鳥獣害対策用の小型無人航空機。

【請求項４】

機体の同一平面上で十字を形成する腕部が延設された筐体の中に、モーターと、同モーターに電力を供給するバッテリーと、同腕部の各々の先端側に回転自在に取り付けられ、前記モーターから動力を得て回転する回転翼と、機体の動作が自動制御されるように、少なくとも、自動離着陸機能と、高度設定機能と、飛行させる区域内の境界点座標値を設定することで飛行区域の位置情報を入手可能なＧＰＳ受信機能と有する自動制御手段とが格納された本体部と、

前記本体部の表面に形成され、光を反射する反射面と、

前記本体部の前記腕部の下部に取着された帯状物であり、光を反射する反射テープと、

前記本体部の下部に取着され、中央部に目玉模様が表されたシートとを備える

鳥獣害対策用の小型無人航空機。

【考案の詳細な説明】

【技術分野】

【０００１】

本考案は、鳥獣害対策用の小型無人航空機に関する。更に詳しくは、畑等を荒ら

す鳥獣について、畑等の上空における飛行による動きや飛翔音、あるいは光等の威嚇手段を用いて威嚇し、撃退できるものに関する。

【背景技術】

【０００２】

従来から、田畑における鳥獣害対策用の様々な器具が考案されている。例えば、下記非特許文献に記載されたような、案山子のような人間を模したもの、あるいは、バルーンや反射光による忌避テープ、定期的に音を放つ機器等が挙げられる。

【０００３】

非特許文献１記載のバルーンは、丸形のバルーンに目玉模様が印刷されたものであり、特に鳥に対して視覚的に刺激を与えて威嚇するものである。非特許文献２記載の人形は、いわゆる案山子や着衣させたマネキン人形等であって、人間が居る体にして、その存在が威嚇となって鳥獣を追い払うものである。

【先行技術文献】

【非特許文献】

【０００４】

【非特許文献１】鳥害対策用品シリーズ　目玉風船（鳥追いバルーン）
http://item.rakuten.co.jp/plusys7022/780/

【０００５】

【非特許文献２】農林水産省　第３章　被害対策の取組事例　第６６頁
（１）カラス対策、　３）対策の方法
http://www.maff.go.jp/j/seisan/tyozyu/higai/h_manual/h20_03a/pdf/tori-data3.pdf

【考案の概要】

【考案が解決しようとする課題】

【０００６】

上記各非特許文献に記載された器具は、鳥獣が嫌う刺激を与えて追い払うものであり、鳥獣を常時観察する人手が不要という利点があるが、同じ刺激を鳥に与え続けると、鳥はその刺激に慣れてしまい、撃退効果が薄れることが多い。

【０００７】

本考案は以上の点に鑑みて創案されたものであって、畑等を荒らす鳥獣に対し、

畑等の上空における飛行による動きや飛翔音、あるいは光等の威嚇手段を用いて威

嚇し、撃退できるものを提供することを目的とするものである。

【課題を解決するための手段】

【０００８】

上記の目的を達成するために本考案の鳥獣害対策用の小型無人航空機は、動力手段と、同動力手段を動かす動力源と、同動力手段から動力を得る推進手段と、機体の動作を制御する遠隔操作手段あるいは自動制御手段とを有する本体部と、
前記本体部に設けられた、鳥獣を威嚇するための威嚇手段とを備える。

【０００９】

ここで、小型無人航空機は、動力手段、動力源、推進手段を有する本体部を備えることによって飛行可能であり、その動きや飛翔音で鳥獣を威嚇することができる。

【００１０】

また、小型無人航空機は、遠隔操作手段あるいは自動制御手段によって、機体の動きを手動あるいはプログラムによる自動制御によって操作することができる。

【００１１】

更に、鳥獣を威嚇するための威嚇手段によって、前述の飛行による動きや飛翔音に加えて、威嚇手段でも鳥獣に刺激が与えられるので、撃退効果が向上する。加えて、機体の動きと威嚇手段が重畳することで、鳥獣が刺激に慣れにくくなることも期待される。

【００１２】

鳥獣を威嚇するための威嚇手段としては、以下のような鳥獣の視覚あるいは聴覚を刺激するものが挙げられる。

なお、本明細書および本実用新案登録請求の範囲でいう鳥獣とは、田畑、果樹園等の農作物に食害を与える動物であり、例えば、スズメ、カラス、ヒヨドリ、ムクドリ等の鳥類、イノシシ、サル、シカ、タヌキ、ハクビシン、アナグマ等の獣類が挙げられる。

【００１３】

威嚇手段が鳥獣の視覚を刺激するものである場合は、例えば、光沢、色彩あるいは模様、または部材の動き、またはこれらの組み合わせ等であって鳥が嫌うものを

付加することによって、小型無人航空機の飛行による動きや飛翔音のみならず、視覚的な威嚇手段でも鳥獣に刺激が与えられるので、撃退効果が更に向上する。加えて、機体の動きと視覚的な威嚇手段が重畳することで、鳥獣が刺激に慣れてしまうことを防止することも期待される。

【００１４】

威嚇手段が鳥獣の聴覚を刺激するものである場合は、例えば、破裂音、天敵の鳴き声、対象鳥獣が仲間に発する警戒音、鳥獣が苦手とする音域の音（好ましくは人間が聞き取れない音域）またはこれらの組み合わせ等であって鳥が嫌うものを発音することによって、小型無人航空機の飛行による動きや飛翔音のみならず、聴覚的な威嚇手段でも鳥獣に刺激が与えられるので、視覚的な威嚇手段のみの場合よりも撃退効果が更に向上する。加えて、機体の動きと聴覚的な威嚇手段が重畳することで、鳥獣が刺激に慣れてしまうことを防止することも期待される。

【００１５】

前記推進手段が回転翼である場合は、垂直方向への立体的な機体動作が可能となるので、機体の動きのバリエーションが増え、ひいては、その動きでも鳥獣を威嚇することができ、かつ、機体の動きのバリエーションの豊富さゆえに、鳥獣が刺激に慣れてしまうことを防止することも期待される。また、垂直離着陸が可能になるので、離着陸のための場所が狭くて済む。

【００１６】

前記本体部の自動制御手段が、自動離着陸機能と、高度設定機能と、飛行させる区域内の境界点座標値を設定することで飛行区域の位置情報を入手可能なＧＰＳ受信機能と有するものである場合は、例えば、自動離着陸機能の時間差設定機能を利用することで、定期的に小型無人航空機が飛行するので、人が常時監視する必要がなくなる。また、高度設定機能、ＧＰＳ受信機能を有しているので、自動で離着陸した後の小型無人航空機を人が操作する必要がなくなる。

【００１７】

上記の目的を達成するために本考案の鳥獣害対策用の小型無人航空機は、機体の同一平面上で十字を形成する腕部が延設された筐体の中に、モーターと、同モーターに電力を供給するバッテリーと、同腕部の各々の先端側に回転自在に取り付けられ、前記モーターから動力を得て回転する回転翼と、機体の動作が自動制御されるように、少なくとも、自動離着陸機能と、高度設定機能と、飛行させる区域内の境界点座標値を設定することで飛行区域の位置情報を入手可能なＧＰＳ受信機能と

有する自動制御手段とが格納された本体部と、前記本体部の表面に形成され、光を反射する反射面と、前記本体部の前記腕部の下部に取着された帯状物であり、光を反射する反射テープと、前記本体部の下部に取着され、中央部に目玉模様が表されたシートとを備える。

【００１８】

ここで、小型無人航空機は、モーター、バッテリーおよび回転翼を備えることにより、垂直離着陸を行うことができると共に、水平方向のみならず高さ方向への機体動作も可能となる。なお、動力手段がモーターであるため、静かに運用できる。

【００１９】

また、小型無人航空機は、機体制御手段として、自動離着陸機能、高度設定機能を有しているので、設定された時間毎に自動運転されて鳥獣への威嚇行動を行うことができ、人が、鳥獣を常時観察し、小型無人航空機をその都度運用する必要が無くなる。

【００２０】

加えて、小型無人航空機は、機体制御手段として、ＧＰＳ受信機能を有しているので、境界点座標値を設定した範囲内（例えば、自分の所有する農地の上空のみ）での航路設定ができ、ナビゲーション機能を利用して自動運転で飛行させることもできるので、仮に小型無人航空機が落下しても、他人の身体や財産（例えば農作物や建物）に危害あるいは損害を与える可能性を低減することができる。

【００２１】

反射面は、太陽光など受けた光を反射するものであり、鳥獣を視覚的に威嚇する。

反射テープは、光を反射して煌めき、また、はためくことで、鳥獣を視覚的に威嚇する。また、反射テープは、飛行時に風切り音を立てるので、鳥獣を聴覚的に威嚇する。

シートは、中央部に表された目玉模様により、特に機体下方にいる鳥獣を視覚的に威嚇する。

【考案の効果】

【００２２】

本考案の鳥獣害対策用の小型無人航空機によれば、畑等を荒らす鳥獣について、畑等の上空における飛行による動きや飛翔音、あるいは光等の威嚇手段を用いて威

嚇し、撃退できる。

【図面の簡単な説明】

【００２３】

【図１】本考案に係る鳥獣害対策用の小型無人航空機の第１実施形態を示す斜視説明図である。

【図２】図１に示す鳥獣害対策用の小型無人航空機の変形例を示す斜め下方から見た斜視
説明図である。

【図３】図２に示す鳥獣害対策用の小型無人航空機の高さ方向の動作の一例を示す使用状態説明図である。

【図４】図２に示す鳥獣害対策用の小型無人航空機の水平方向の動作の一例を示す使用状態説明図である。

【図５】本考案に係る鳥獣害対策用の小型無人航空機の第２実施形態を示す斜視説明図である。

【図６】本考案に係る鳥獣害対策用の小型無人航空機の第３実施形態を示す斜視説明図である。

【考案を実施するための形態】

【００２４】

図１乃至図６を参照して、本考案の実施の形態を更に詳細に説明する。なお、各図における符号は、煩雑さを軽減し理解を容易にする範囲内で付している。また、以下で述べる用語「回転翼」「プロペラ」は先に述べた「推進手段」と、用語「反射面」「吊り下げ体」「吹き流し体」「スピーカー部」は先に述べた「威嚇手段」と、それぞれ同等の意味で使用している。

【００２５】

〔第１実施形態〕

図１に示す鳥獣害対策用の小型無人航空機Ｐ１は、本体部１０と、本体部１０に設けられた反射面１３０とを備える。各部については、以下詳述する。

【００２６】

本体部１０は、軽量かつ剛性を有する樹脂（本実施形態ではグラスファイバーで強化されたＡＢＳ樹脂）で中空に形成され、使用時において水平方向となる方向へ

十字状の腕部が延設された筐体の中に、動力手段（本実施形態ではモーター、図示省略）と、動力手段を動かす動力源（本実施形態ではリチウムイオンポリマー二次電池、図示省略）と、機体の動作を制御する遠隔操作手段および自動制御手段（本実施形態では、プロセッサ、通信装置、ＧＰＳ（Global Positioning System 全地球測位システム）、磁力計、ジャイロスコープ、加速度計、超音波高度センサー、気圧計等。図示省略）が格納されている。また、本体部１０は、本体部下面に、下方に垂下し、かつ、間隔を開けて取り付けられた２本の脚部１２を有している。

【００２７】

回転翼１１は、本体部１０の先端側の各々に回転自在に取り付けられ（合計４つ）、動力手段から動力を得て、機体を飛行可能とする構造である。つまり、本実施形態における小型無人航空機Ｐ１は、複数の回転翼１１を有する回転翼機（いわゆるマルチコプター）である。

【００２８】

反射面１３０は、ホログラム加工のカッティングシートであり、本体部１０の表面に貼り付けて形成してある。反射面１３０は、太陽光など受けた光を多色で強く反射するものであり、鳥獣の視覚を刺激する威嚇手段にあたる。

【００２９】

なお、本体部１０の下面側は陰となって太陽光が直接当たりにくいので、本体部１０下面を照射するように配置したＬＥＤライトを本体部１０の下方に設けてもよい。この場合、小型無人航空機Ｐ１の下方にいる鳥獣についても、反射面１３０から生じる反射光による威嚇ができ、撃退効果が向上する。

【００３０】

本実施の形態においては、本体部１０の筐体部分の材質は、グラスファイバーで強化されたＡＢＳ樹脂を採用しているが、これに限定するものではなく、例えば、

他の種類の合成樹脂、炭素繊維、アルミニウム合金、マグネシウム合金等の各種金属等、またはこれらの組み合わせであってもよく、軽量かつ剛性を有するものであることが好ましい。

【００３１】

本実施の形態においては、本体部１０の形状が十字状であるが、これに限定するものではなく、例えば、円盤形や、多角形の板状体または多角形の錐体等、様々な形状でつくることができる。また、本実施の形態においては、回転翼１１は、４つ取り付けられているが、これに限定するものではなく、例えば、本体部の形状や要求される性能等に合わせて、数および配置を適宜設定することができる。

【００３２】

本実施の形態においては、静粛性の観点から動力手段としてモーターを採用しているが、これに限定するものではなく、例えば、エンジン等の公知の動力手段を除外するものではない。また、本実施の形態においては、動力源はリチウムイオンポリマー二次電池であるが、これに限定するものではなく、例えば、他の種類のバッテリーであってもよく、前述のように動力手段がエンジンであれば、ラジコン燃料（メチルアルコール、潤滑油、ニトロメタンの混合燃料)、ガソリン等の液体燃料であってもよい。

【００３３】

本実施の形態においては、鳥獣に対する威嚇手段として反射面１３０（鳥獣の視覚を刺激するもの）を採用しているが、これに限定するものではなく、例えば、本体部自体に反射効果がある塗料を塗布してもよいし、また、鳥獣が嫌う模様（例えば、タカやハヤブサの絵や羽の模様）や色彩を採用してもよい。また、本体部１０のみならず、回転翼１１にも鳥獣が嫌う模様や色彩を採用してもよい。更に、本体部や回転翼に、鳥獣の視覚を刺激すべく、点滅機能を有するＬＥＤライト等の発光部を設けてもよい。

【００３４】

（変形例）

図２に示す小型無人航空機Ｐ２は、第１実施形態の小型無人航空機Ｐ１の変形例であり、小型無人航空機Ｐ１と同様に、鳥獣の視覚を刺激する威嚇手段を採用した

ものである。

【００３５】

小型無人航空機Ｐ２は、鳥獣に対する威嚇手段として、本体部１０下面側に取り付けられた目玉状の吊り下げ体１３１、メタリックカラーの吹き流し体１３２を備えている。なお、吊り下げ体１３１と吹き流し体１３２を備えている点以外は、小型無人航空機Ｐ１と同様の構造であるため、説明を省略する。

【００３６】

（作　用）

図３および図４を参照して、小型無人航空機Ｐ２の作用について説明する。なお、図３および図４で示す小型無人航空機Ｐ２は、その作用が吊り下げ体１３１と吹き流し体１３２によるもの以外は、小型無人航空機Ｐ１の作用と同じであるため、小型無人航空機Ｐ１の作用に関する説明は省略する。

【００３７】

田畑等に小型無人航空機Ｐ２を持ち込み、田畑等の上空に飛ばして、作物を狙う鳥獣（例えば、カラス５１、イノシシ５２、猿５３等）を撃退する。このとき、小型無人航空機Ｐ２は、無線送信機（プロポーショナルシステム）によって手動操作にもできるが、携帯端末にインストールされた機体操作アプリケーションを使用して、自動制御運転とすることもできる。小型無人航空機Ｐ２を自動制御運転にすることで、人が、鳥獣を常時観察し、小型無人航空機Ｐ２を操作する必要が無くなる。

【００３８】

小型無人航空機Ｐ２は、マルチコプターであるため、垂直離着陸が可能であり、離着陸のための場所が狭くて済むと共に、飛行領域における上下方向（高低方向）への立体的な機体動作が可能となる（図３参照）。このため、一定高度を飛行するのみ可能なものよりも、機体の複雑な動作を行うことができ、これによって鳥獣への威嚇効果が高まる。

【００３９】

小型無人航空機Ｐ２は、一般的なマルチコプターと同様に、本体部に設けた複数の回転翼を同時にバランスよく回転させることによって飛行し、上昇あるいは下降が回転翼の回転数の増減によって行われ、前進あるいは後進の水平方向への移動が回転翼の回転数の増減によって機体を傾けることで進むようになっている。なお、回転翼は、固定ピッチの物が採用されることが多く、右回りと左回りのものが交互に配置されることで、回転の反作用を打ち消しあっている。

【００４０】

また、自動制御運転であれば、ジャイロスコープ、超音波高度センサー等の各種センサーと、プロセッサにインストールされた運転制御プログラムとが協働し、設定された、あるいは観測された情報に基づいて、所定の高度等の範囲内での立体的な機体動作が行われる。

【００４１】

このとき、小型無人航空機Ｐ２は、その動きや飛翔音、風切り音で、鳥獣を視覚的、聴覚的に威嚇することができ、加えて、本体部１０からの反射光、吊り下げ体１３１あるいは吹き流し体１３２の光、色や動きでも鳥獣に視覚的刺激が与えられ、特に、吊り下げ体１３１は、飛行時に風切り音が立って鳥獣に聴覚的刺激も与えられるので、機体からの反射光のみの場合よりも撃退効果が向上する。更に、機体の動きと威嚇手段が重畳することで、鳥獣が刺激に慣れにくくなることも期待される。

【００４２】

加えて、小型無人航空機Ｐ２は、ＧＰＳを備えることで、境界点座標値を設定した範囲内（例えば、自分の所有する農地Ｆの上空のみ）での飛行を自動運転で行うこともできる（図４参照）。これにより、仮に小型無人航空機Ｐ２が落下しても、他人の身体や財産（例えば農作物や建物）に害を加える危険性を抑えることができる。

【００４３】

更に、小型無人航空機Ｐ２は、自動制御手段における時間差設定機能を用いて、自動離着陸が可能であり、設定された時間毎に自動運転され、鳥獣への威嚇行動を行うことができるので、人が、鳥獣を常時観察し、小型無人航空機Ｐ２をその都度運用する必要が無くなる。

【００４４】

〔第２実施形態〕

図５に示す小型無人航空機Ｐ３は、本考案の他の実施形態である。

小型無人航空機Ｐ３は、鳥獣に対する威嚇手段として、鳥獣の聴覚を刺激する音声データ（銃器の発砲音のような破裂音、天敵の鳴き声、対象鳥獣が仲間に発する警戒音等の複数の音声）を記録した音声記録部（図示省略）と、本体部１０下面側に取り付けられ、音声記録部からの音声を出力するスピーカー部１４と、を備えている。

【００４５】

スピーカー部１４は複数のスピーカーを有し、操縦者からの操作で、あるいは自動制御手段からの定期的なコマンドにより、鳥獣の聴覚を刺激する音を発する。

なお、小型無人航空機Ｐ３は、スピーカー部１４を備えている点以外は、小型無人航空機Ｐ１と同様の構造であるため、その説明を省略する。

【００４６】

小型無人航空機Ｐ３によれば、機体の動きや反射光等による視覚的な刺激に加え、スピーカー部１４から出力される様々な音声によって、鳥獣に聴覚的な刺激が与えられ、視覚的な刺激のみが与えられる場合よりも、撃退効果が向上する。更に、機体の動きと音声による威嚇手段が重畳することで、鳥獣が刺激に慣れにくくなることも期待される。

【００４７】

〔第３実施形態〕

図６に示す小型無人航空機Ｐ４は、固定翼機であり、その機体表面にホログラム加工のカッティングシートを貼り付けてなる反射面１３０ａを有し、主翼、水平尾翼の先端に、吹き流し体１３２が取り付けられている。軽量かつ剛性を有する樹脂で中空に形成された機体の内部には、動力手段（本実施形態ではモーター、図示省略）と、動力手段を動かす動力源（本実施形態ではリチウムイオンポリマー二次電池、図示省略）と、機体の動作を制御する遠隔操作手段および自動制御手段（本実施形態では、プロセッサ、通信装置、ＧＰＳ（Global Positioning System 全地球測位システム）、磁力計、ジャイロスコープ、加速度計、超音波高度センサー、気圧計等。図示省略）が格納されており、機体の上部にプロペラ１１０を１つ有している。

【００４８】

この構成によれば、本実施形態における小型無人航空機Ｐ４では、小型無人航空機Ｐ１～Ｐ３のような急激な立体的な機体動作は行いにくいが、プロペラ１１０が１つなので、バッテリー消費量が少なく、複数の回転翼を有するような機体と比較して長時間の運用が可能となる。

【００４９】

また、小型無人航空機Ｐ４は、小型無人航空機Ｐ２と同様に、機体の動き、機体からの反射光および吹き流し体１３２による視覚的刺激、飛行時の風切り音による聴覚的刺激によって、鳥獣を視覚的、聴覚的に威嚇することができ、更に、機体の動きと威嚇手段が重畳することで、鳥獣が刺激に慣れにくくなることも期待される。

【００５０】

なお、第１から第４の実施形態の小型無人航空機Ｐ１～Ｐ４は、その翼端や機体に風切り音や笛鳴り現象を生じさせる部分を形成し、音による威嚇効果を高めるようにしてもよい。

【００５１】

このように、第１実施形態（変形例を含む）から第３実施形態の小型無人航空機Ｐ１～Ｐ４によれば、畑等を荒らす鳥獣５１、５２、５３に対し、農地Ｆのような畑等の上空における飛行による動きや飛翔音、あるいは光、音の威嚇手段を用いて威嚇し、追い払うことができる。

【００５２】

なお、本実用新案登録請求の範囲および本明細書で使用している用語と表現は、あくまでも説明上のものであって、なんら限定的なものではなく、本実用新案登録請求の範囲および本明細書に記述された特徴およびその一部と等価の用語や表現を除外する意図はない。また、本考案の技術思想の範囲内で、種々の変形態様が可能であるということは言うまでもない。

【符号の説明】

【００５３】

Ｐ１、Ｐ２、Ｐ３、Ｐ４　　小型無人航空機

１０　　本体部

１１　　回転翼

１１０　　プロペラ

１２　　脚部

１３０、１３０ａ　　反射面

１３１　　吊り下げ体

１３２　　吹き流し体

１４　　スピーカー部

５１　　カラス

５２　　イノシシ

５３　　猿

Ｆ　　農地

図１

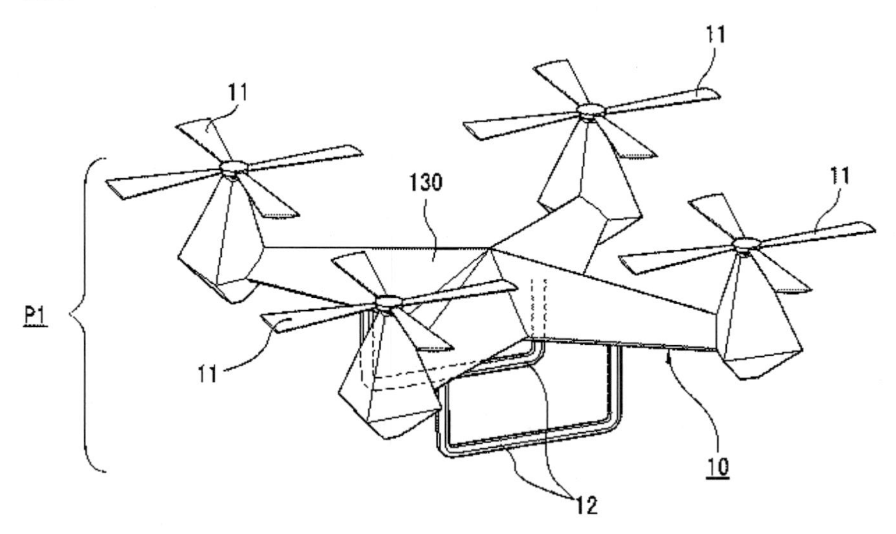

図２

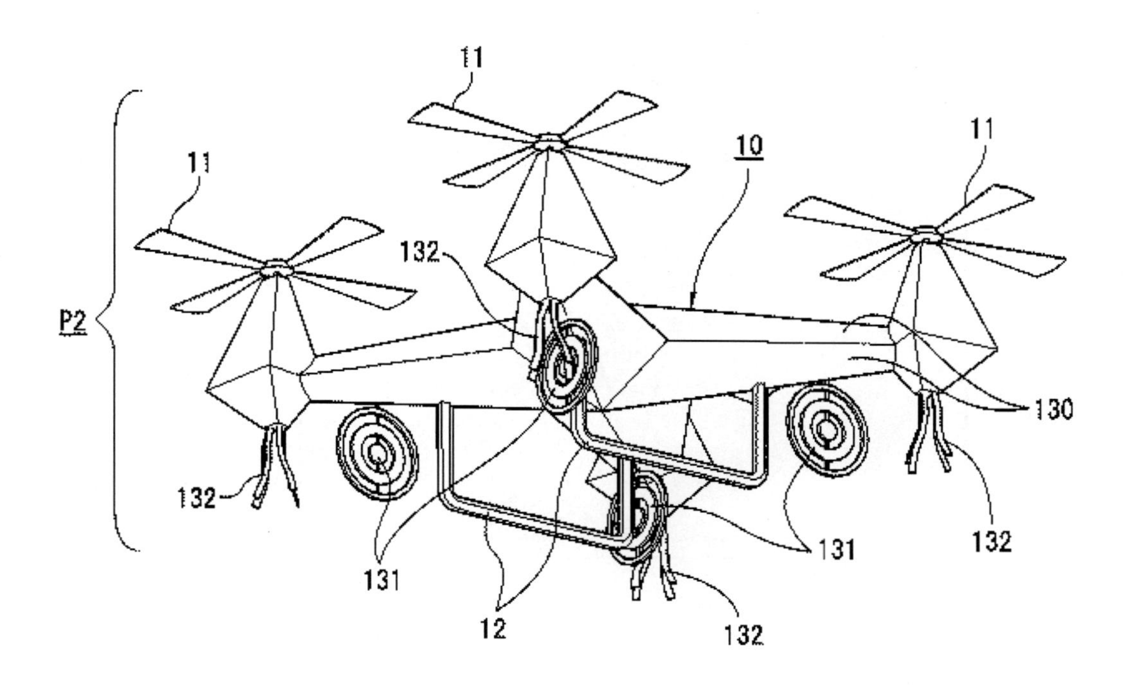

図３

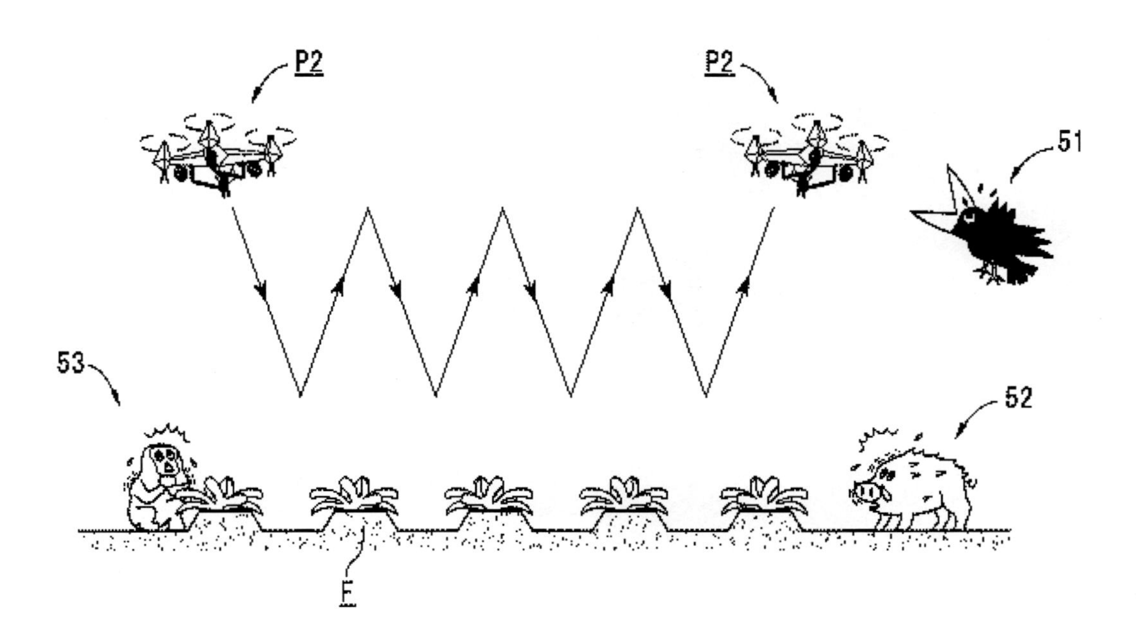

図４

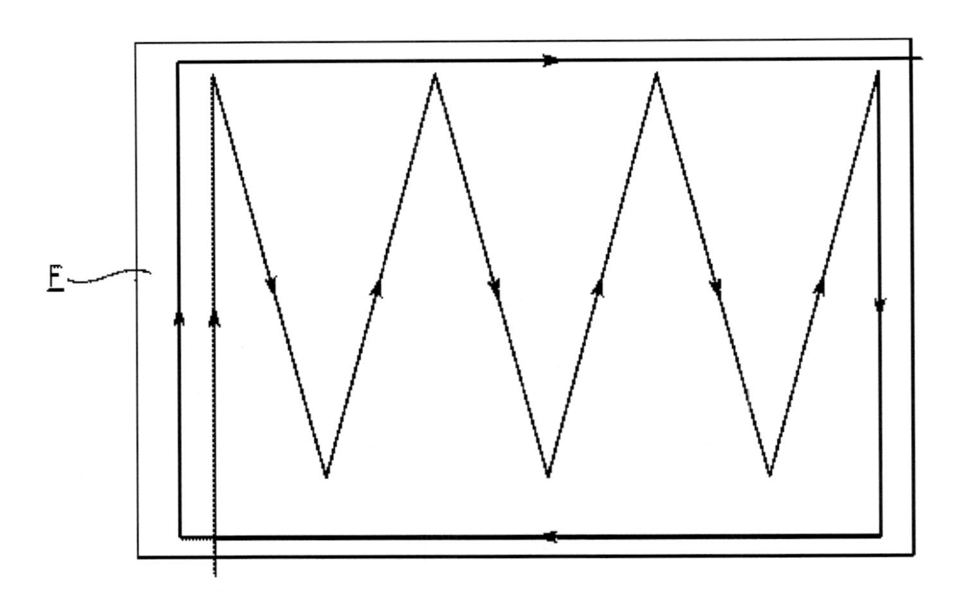

図５

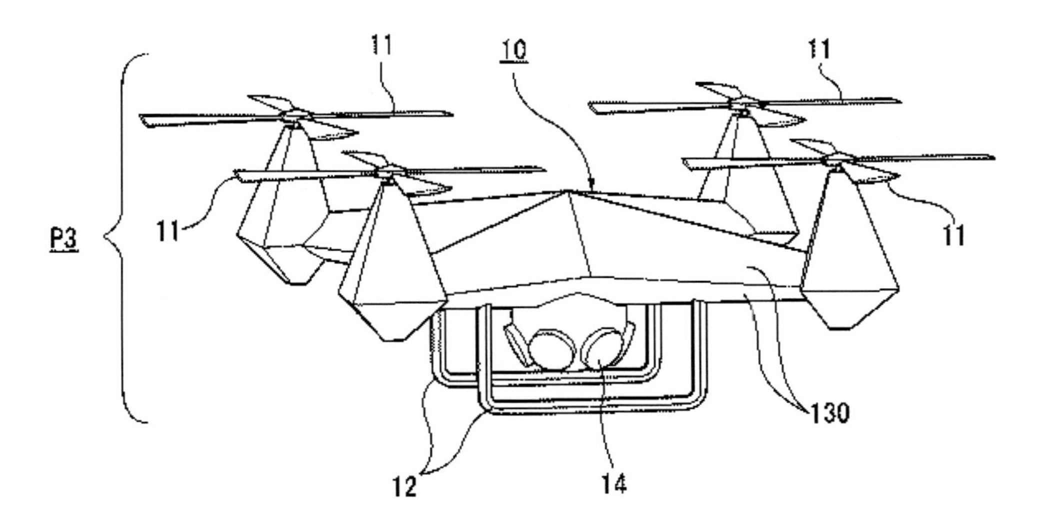

図 6

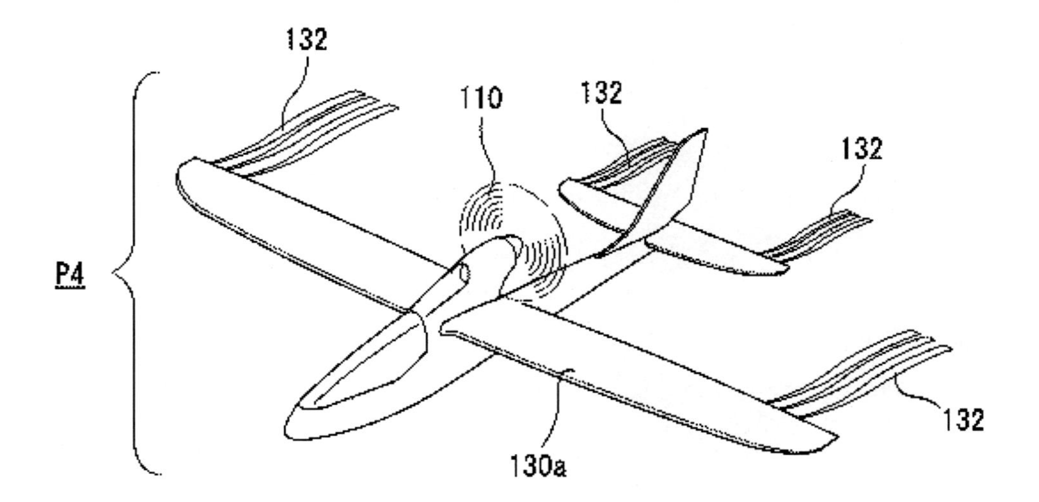

(3)

Patent journal English

[Claim 1]

A body part which has a power means, the source of power to which the power means is moved, a propulsion means which obtains power from the power means, and a remote operation means or an automatic controlling means which controls operation of a body,

It has an intimidation means for threatening wildlife provided by the aforementioned body part.

A small unmanned aircraft for a measure against wildlife damage.

[Claim 2]

The aforementioned propulsion means is a rotary wing.

A small unmanned aircraft for the measure against wildlife damage according to claim 1.

[Claim 3]

An automatic controlling means of the aforementioned body part has the position information on a flight zone with an available GPS receiving function by setting a boundary point coordinate value in a zone to fly to an automatic taking-off-and-landing function and an altitude-thresh function.

A small unmanned aircraft for the measure against wildlife damage according to claim 1 or 2.

[Claim 4]

In an installed housing, an arm which forms a cross joint on the same plane of a body A motor, So that automatic control of the operation of a body may be carried out to a battery which supplies electric power to the motor, and a rotary wing which is rotatably attached to each tip side of the arm, obtains power from the aforementioned motor, and it rotates. A body part in which an automatic taking-off-and-landing function, an altitude-thresh function, and an automatic controlling means that has the position information on a flight zone with an available GPS receiving function by setting up a boundary point coordinate value in a zone to fly were stored at least,

A reflective surface in which it is formed in the surface of the aforementioned body part, and light is reflected,

A reflection type which is the strip-like thing attached in the lower part of the aforementioned arm of the aforementioned body part, and reflects light,
It is attached in the lower part of the aforementioned body part, and has a sheet by which eyeball patterns were expressed with a central part.
A small unmanned aircraft for a measure against wildlife damage.

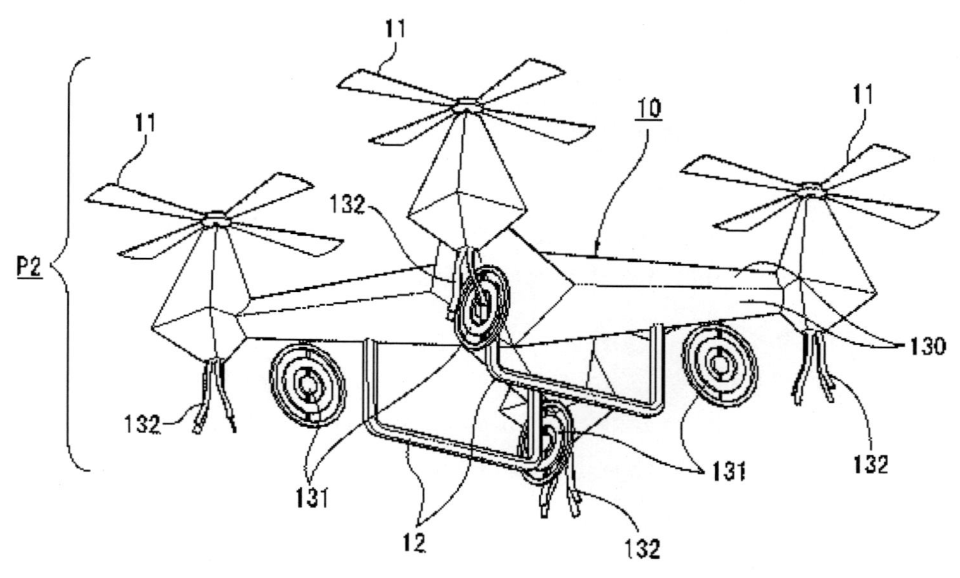

Representative drawing

DETAILED DESCRIPTION

[Detailed explanation of the device]

[Field of the Invention]

[0001]

This design is related with the small unmanned aircraft for the measure against wildlife damage. It threatens in detail about the wildlife which damages a field etc. using intimidation means, such as a motion by the flight in the skies, such as a field, a flight sound or light, and is related with what can be repulsed.

[Background of the Invention]

[0002]

From the former, various instruments for a wildlife damage measure in fields are devised. For example, a thing which was described in the following nonpatent literature and which imitated human being like a scarecrow or the evasion tape by the balloon or a reflected light, the apparatus emit a sound periodically, etc. are mentioned.

[0003]

Eyeball patterns are printed at a round balloon, and the balloon of nonpatent literature 1 description gives and threatens a stimulus visually especially to a bird. The doll of nonpatent literature 2 description is what is called a scarecrow, the mannequin dummy made to wear one's clothes, etc., and it is made the body in which human being is, the existence serves as intimidation, and it drives off wildlife.

[Citation list]

[Nonpatent literature]

[0004]

[Nonpatent literature 1] Damage-by-birds measure article series Eyeball balloon (****** balloon) http://item.rakuten.co.jp/plusys7022/780/

[0005]

[Nonpatent literature 2] Ministry of Agriculture, Forestry, and Fisheries Chapter 3 Measure example of the measure against damage Method http: of measure against 66th page (1) crow, and threemeasures

//www.maff.go.jp/j/seisan/tyozyu/higai/h_manual/h20_03 a/pdf/tori-data3.pdf

[The outline of a device]

[Problem(s) to be Solved by the Device]

[0006]

Although the instrument described in each above-mentioned nonpatent literature gives and drives off the stimulus which wildlife dislikes and has the advantage that the help who always observes wildlife is unnecessary, if it continues giving the same stimulus to a bird, a bird will get used to the stimulus and the repulse effect will fade in many cases.

[0007]

This design is originated in view of the above point, and it aims at threatening using intimidation means, such as a motion by the flight in the skies, such as a field, a flight sound or light, and providing what can be repulsed to the wildlife which damages a field etc.

[Means for solving problem]

[0008]

The body part which has the source of power to which the small unmanned aircraft for the measure against wildlife damage of this design moves a power means and the power means, a propulsion means which obtains power from the power means, and the remote operation means or automatic controlling means which controls operation of the body in order to achieve the above-mentioned object,

It has the intimidation means for threatening wildlife provided by the aforementioned body part.

[0009]

Here, the small unmanned aircraft can fly by having a body part which has a power means, the source of power, and a propulsion means, and can threaten wildlife to the motion and flight sound.

[0010]

A small unmanned aircraft can be operated by the automatic control according a motion of the body to hand control or a program by the remote operation means or an automatic controlling means.

[0011]

Since a stimulus is given to wildlife also by an intimidation means by the intimidation means for threatening wildlife in addition to the motion and flight sound by the above-mentioned flight, the repulse effect improves. In addition, a motion and intimidation means of the body are expected by a superimposed thing [carrying out] that wildlife becomes difficult to get used to a stimulus.

[0012]

As an intimidation means for threatening wildlife, what stimulates the vision or the acoustic sense of the following wildlife is mentioned.

The wildlife as used in the field of this Description and these Claims is an animal which does a vermin damage to agricultural products, such as fields and an orchard, for example, brutes, such as birds, such as a sparrow, a crow, a brown-eared bulbul, and MUKUDORI, a wild boar, an ape, a deer, a raccoon dog, a masked palm civet, and a badger, are mentioned.

[0013]

When an intimidation means is what stimulates the vision of wildlife, For example, since a stimulus is given to wildlife not only by the motion by the flight of a small unmanned aircraft, or a flight sound but by a visual intimidation means by adding what is motions of gloss, color, a pattern, or a component or such combination, and a bird dislikes, the repulse effect further improves. In addition, it is also expected that a motion of the body and a visual intimidation means will prevent wildlife from getting used to a stimulus by a superimposed thing [carrying out].

[0014]

When an intimidation means is what stimulates the acoustic sense of wildlife, For example, by pronouncing what is sounds (compass which human being cannot catch preferably) or such combination of the compass which an explosive sound, the cry of a natural enemy, and object wildlife make poor at the watch sound and wildlife which are emitted to an associate, etc., and a bird dislikes, Since a stimulus is given to wildlife not only by the motion by the flight of a small unmanned aircraft, or a flight sound but by an auditory intimidation means, the repulse effect further improves rather than the case of only a visual intimidation means. In addition, it is also expected that a motion of the body and an auditory intimidation means will prevent wildlife from getting used to a stimulus by a superimposed thing [carrying out].

[0015]

Since the three-dimensional body operation to vertical direction is attained when the aforementioned propulsion means is a rotary wing, the variation of a motion of the body increases, and if it pulls, Preventing the abundance, therefore wildlife of a variation of a motion of the body from being able to threaten wildlife also by the motion and getting used to a stimulus is also expected. Since vertical takeoff and

landing become possible, the place for taking off and landing is narrow, and ends.
[0016]
The automatic controlling means of the aforementioned body part An automatic taking-off-and-landing function and an altitude-thresh function, Since it is using the time lag setting up function of an automatic taking-off-and-landing function and a small unmanned aircraft flies periodically, for example when it is what has the position information on a flight zone with an available GPS receiving function by setting up the boundary point coordinate value in the zone to fly, it becomes unnecessary for people to monitor continuously. Since it has an altitude-thresh function and a GPS receiving function, it becomes unnecessary for people to operate the small unmanned aircraft after making take-off and landing automatically.
[0017]
In order to achieve the above-mentioned object, a small unmanned aircraft for a measure against wildlife damage of this design is provided with the following.
In a housing in which an arm which forms a cross joint on the same plane of a body was installed, it is a motor.
A battery which supplies electric power to the motor.
A rotary wing which is rotatably attached to each tip side of the arm, obtains power from the aforementioned motor, and it rotates.
At least so that automatic control of the operation of the body may be carried out An automatic taking-off-and-landing function, A body part in which an altitude-thresh function and an automatic controlling means which has the position information on a flight zone with an available GPS receiving function by setting up a boundary point coordinate value in a zone to fly were stored, A sheet by which it was formed in the surface of the aforementioned body part, and was attached in a reflective surface in which light is reflected, a reflection type which is the strip-like things attached in the lower part of the aforementioned arm of the aforementioned body part, and reflects light, and the lower part of the aforementioned body part, and eyeball patterns were expressed with a central part.
[0018]
Here, the small unmanned aircraft can perform vertical takeoff and landing by having a motor, a battery, and a rotary wing, and the body operation of it to a height direction is also attained failing in a horizontal chisel. Since a power means

is a motor, it can apply calmly.

[0019]

Since the small unmanned aircraft has an automatic taking-off-and-landing function and an altitude-thresh function as a body control means, the set-up necessity of it that it is operated automatically for every time and can perform threat behavior to wildlife, and people always observe wildlife and employ a small unmanned aircraft each time is lost.

[0020]

In addition, since the small unmanned aircraft has a GPS receiving function as a body control means, Since route setting out within the limits (for example, only sky of the farmland which he owns) which set up the boundary point coordinate value can be performed and it can also be made to fly by automatic operation using a navigation function, Even if a small unmanned aircraft falls, a possibility of doing harm or damage to others' body and property (for example, agricultural products and a building) can be reduced.

[0021]

A reflective surface reflects received lights, such as sunlight, and threatens wildlife visually.

A reflection type reflects light, are ****** and fluttering and threaten wildlife visually. Since a reflection type raises a whizzing sound at the time of a flight, it threatens wildlife auditorily.

A sheet threatens visually the wildlife which is especially in a body lower part according to the eyeball patterns denoted by the central part.

[Effect of the Device]

[0022]

According to the small unmanned aircraft for the measure against wildlife damage of this design, it can threaten and repulse about the wildlife which damages a field etc. using intimidation means, such as a motion by the flight in the skies, such as a field, a flight sound or light.

[Brief Description of the Drawings]

[0023]

[Drawing 1]It is a strabism explanatory view showing a 1st embodiment of the small unmanned aircraft for the measure against wildlife damage concerning this design.

[Drawing 2]Strabism seen from the slanting lower part which shows the modification of the small unmanned aircraft for the measure against wildlife damage shown in Fig.1

It is an explanatory view.

[Drawing 3]It is a busy condition explanatory view showing an example of operation of the height direction of the small unmanned aircraft for the measure against wildlife damage shown in Fig.2.

[Drawing 4]It is a busy condition explanatory view showing an example of horizontal operation of the small unmanned aircraft for the measure against wildlife damage shown in Fig.2.

[Drawing 5]It is a strabism explanatory view showing a 2nd embodiment of the small unmanned aircraft for the measure against wildlife damage concerning this design.

[Drawing 6]It is a strabism explanatory view showing a 3rd embodiment of the small unmanned aircraft for the measure against wildlife damage concerning this design.

[The form for devising]

[0024]

With reference to Fig.1 thru/or Fig.6, the embodiment of this design is described still in detail. The code in each figure is attached within limits which understand easily by reducing complicatedness. The term "rotary wing" described below and the "propeller" are using the "propulsion means" expressed previously, a term "reflective surface" and a "hanging object", the "streamer object", and the "loudspeaker part" in the meaning respectively equivalent to the "intimidation means" described previously.

[0025]

[A 1st embodiment]

The small unmanned aircraft P1 for a measure against wildlife damage shown in Fig.1 is provided with the following.

Body part 10.

The reflective surface 130 provided by the body part 10.

Each part is explained in full detail below.

[0026]

The body part 10 is formed in midair by the resin (ABS plastics strengthened with this embodiment of glass fiber) which has a light weight and rigidity, In the installed housing, a cross shape arm in the direction which becomes horizontal at the time of use A power means (this embodiment a motor, a graphic display abbreviation), The remote operation means and automatic controlling means (according to this embodiment) which control the source of power (this embodiment a lithium ion polymer secondary cell, a graphic display abbreviation) to which a power means is moved, and operation of the body A processor, a communication apparatus, GPS (Global Positioning System Global Positioning System), a magnetometer, a gyroscope, an accelerometer, an ultrasonic altitude sensor, a manometer, etc. The graphic display abbreviation is stored. The body part 10 has the two legs 12 which hung on the body part lower surface caudad, and opened the interval in it, and were attached to it.

[0027]

The rotary wing 11 is a structure which is rotatably attached to each on the tip side of the body part 10 (a total of four), obtains power from a power means, and enables the flight of the body. That is, the small unmanned aircraft P1 in this embodiment is a rotorcraft (what is called multi-KOPUTA) which has a plurality of rotary wings 11.

[0028]

The reflective surface 130 is a cutting sheet of hologram processing, and it has stuck and formed it in the surface of the body part 10. The reflective surface 130 reflects received lights, such as sunlight, strongly with multiple color, and hits an intimidation means to stimulate the vision of wildlife.

[0029]

the lower surface side of the body part 10 serves as shade, and its sunlight is direct -- per -- being hard -- since -- the LED light arranged so that body part 10 lower surface may be irradiated -- the body part 10 -- caudad -- it may provide . In this case, also about the wildlife in which the small unmanned aircraft P1 is caudad, intimidation by the reflected light which arises from the reflective surface 130 is made, and the repulse effect improves.

[0030]

In this embodiment, although the ABS plastics strengthened with glass fiber are adopted, the material of the housing portion of the body part 10, it is not what is limited to this -- various metal, such as a synthetic resin of other kinds for example, carbon fiber, an aluminum alloy, and a Magnesium alloy, etc. -- or -- these -- combine -- it is preferable that it is what comes out, may exist and has a light weight and rigidity.

[0031]

In this embodiment, although the form of the body part 10 is cross shape, it cannot limit to this and can build with various form, such as a disk form, a polygonal plate or a polygonal cone, for example. In this embodiment, although the four rotary wings 11 are attached, they cannot be limited to this and can set up a number and arrangement as appropriate, for example to compensate for form, performance demanded, etc. of a body part.

[0032]

In this embodiment, although the motor is adopted as a power means in view of quietness, it does not limit to this and does not except publicly known power means, such as an engine, for example. In this embodiment, although the source of power is a lithium ion polymer secondary cell, As long as it may be not the thing to limit to this but a battery of other kinds for example, and a power means is an engine as mentioned above, they may be liquid fuel, such as radio control fuel (methyl alcohol, a lubricating oil, the composite fuel of nitromethane), and gasoline.

[0033]

In this embodiment, although the reflective surface 130 (what stimulates the vision of wildlife) is adopted as an intimidation means against wildlife, Signs (for example, a hawk, the picture of a falcon, and the pattern of feather) that not the thing to limit to this but the coating which has a reflection effect in the body part itself, for example may be applied, and wildlife dislikes, and color may be adopted. Signs that wildlife dislikes, and color may be adopted not only as the body part 10 but as the rotary wing 11. Light-emitting parts, such as a LED light which has a blink function, may be provided that the vision of wildlife should be stimulated to a body part or a rotary wing.

[0034]

(Modification)

The small unmanned aircraft P2 shown in Fig.2 is a modification of the small unmanned aircraft P1 of a 1st embodiment, and an intimidation means to stimulate the vision of wildlife is used for it like the small unmanned aircraft P1.

[0035]

The small unmanned aircraft P2 is provided with the eyeball-like hanging object 131 attached to the body part 10 lower-surface side, and the streamer object 132 of metallic color as an intimidation means against wildlife. Except a point provided with the hanging object 131 and the streamer object 132, since it is the same structure as the small unmanned aircraft P1, a description is omitted.

[0036]

(Work for)

With reference to Fig.3 and Fig.4, it describes about an operation of the small unmanned aircraft P2. Since the small unmanned aircraft P2 shown by Fig.3 and Fig.4 is the same as an operation of the small unmanned aircraft P1 except what the operation hangs and is depended on the body 131 and the streamer object 132, the description about an operation of the small unmanned aircraft P1 omits it.

[0037]

The small unmanned aircraft P2 is carried into fields etc., and it flies over fields etc., and repulses the wildlife (for example, the crow 51, the wild boar 52, the monkey 53, etc.) which aims at crops. Although the small unmanned aircraft P2 is made also into manual operation with a radio transmitter (proportional system) at this time, the body operation application installed in the personal digital assistant can be used, and it can also be considered as automatic control operation. The necessity that people always observe wildlife and operate the small unmanned aircraft P2 by making the small unmanned aircraft P2 into automatic control operation is lost.

[0038]

Since the small unmanned aircraft P2 is multi-KOPUTA, vertical takeoff and landing are possible, the place for taking off and landing is narrow, and ends, and the three-dimensional body operation of it to the vertical direction (the direction of height) in an air traffic window is attained (refer to Fig.3). For this reason, rather than what only has possible flying a constant altitude, complicated operation of the body can be performed and the intimidation effect to wildlife increases by this.

[0039]

The small unmanned aircraft P2 flies by rotating simultaneously a plurality of rotary wings provided to the body part with sufficient balance like general multi-KOPUTA, A rise or descent is performed by the change in the rotational speed of a rotary wing, and movement to the horizontal direction of advance or sternway proceeds by leaning the body by the change in the rotational speed of a rotary wing. The thing of a fixed pitch is adopted in many cases, and it is that the thing of right-handed rotation and left-handed rotation is arranged alternately, and a rotary wing negates a rotational reaction, and suits.

[0040]

If it is automatic control operation, based on the information which various sensors, such as a gyroscope and an ultrasonic altitude sensor, and the operation control program installed in the processor cooperated, and was set up or observed, three-dimensional body operation in within the limits, such as a predetermined altitude, will be performed.

[0041]

At this time, the small unmanned aircrafts P2 are that motion and flight sound, and a whizzing sound, Can threaten wildlife visually and auditorily and In addition, the reflected light from the body part 10, A visual cue is given to wildlife also the light of the hanging object 131 or the streamer object 132, a color, and a motion, and since a whizzing sound stands at the time of a flight and an auditory stimulus is also given to wildlife, the repulse effect of the hanging object [especially] 131 improves rather than the case of only the reflected light from the body. A motion and intimidation means of the body are expected by a superimposed thing [carrying out] that wildlife becomes difficult to get used to a stimulus.

[0042]

In addition, the small unmanned aircraft P2 can also perform the flight within the limits (for example, only sky of the farmland F which he owns) which set up the boundary point coordinate value by having GPS by automatic operation (refer to Fig.4). Even if the small unmanned aircraft P2 falls by this, the danger of inflicting damage on others' body and property (for example, agricultural products and a building) can be suppressed.

[0043]

Since automatic taking off and landing is possible, it is operated automatically for every set-up time using the time lag setting up function in an automatic controlling means and the small unmanned aircraft P2 can perform threat behavior to wildlife, people always observe wildlife and the necessity of it of employing the small unmanned aircraft P2 each time is lost.

[0044]

[A 2nd embodiment]

The small unmanned aircrafts P3 shown in Fig.5 are other embodiments of this design.

the voice data (an explosive sound like the firing sound of small arms --) which stimulates the acoustic sense of wildlife as an intimidation means as opposed to wildlife in the small unmanned aircraft P3 The cry of a natural enemy and object wildlife were attached to the body part 10 lower-surface side with the voice recording part (not shown) which recorded a plurality of sounds, such as a watch sound emitted to an associate, and are provided with the loudspeaker part 14 which outputs the sound from a voice recording part.

[0045]

The loudspeaker part 14 has a plurality of speakers, is operation from a manipulator or emits the sound which stimulates the acoustic sense of wildlife by the periodical command from an automatic controlling means.

Except a point provided with the loudspeaker part 14, since the small unmanned aircraft P3 is the same structure as the small unmanned aircraft P1, it omits the description.

[0046]

According to the small unmanned aircraft P3, the repulse effect improves rather than the case where in addition to the visual stimulus by a motion, a reflected light, etc. of the body an auditory stimulus is given to wildlife by various sounds outputted from the loudspeaker part 14, and only a visual stimulus is given. An intimidation means with a motion and sound of the body is expected by a superimposed thing [carrying out] that wildlife becomes difficult to get used to a stimulus.

[0047]

[A 3rd embodiment]

The small unmanned aircraft P4 shown in Fig.6 is a fixed-wing aircraft, it has the reflective surface 130a which sticks the cutting sheet of hologram processing on the body surface, and the streamer object 132 is attached at the tip of a main plane and a horizontal tail plane. Inside the body formed in midair by the resin which has a light weight and rigidity, A power means (this embodiment a motor, a graphic display abbreviation) and the source of power (this embodiment a lithium ion polymer secondary cell, a graphic display abbreviation) to which a power means is moved, The remote operation means and automatic controlling means which control operation of the body (this embodiment a processor, a communication apparatus, GPS (Global Positioning System Global Positioning System), a magnetometer, a gyroscope, an accelerometer, an ultrasonic altitude sensor, a manometer, etc.) The graphic display abbreviation is stored and it has the one propeller 110 in the upper part of the body.

[0048]

According to this composition, with the small unmanned aircraft P4 in this embodiment, Although it is hard to perform rapid three-dimensional body operation like the small unmanned aircrafts P1-P3, since the number of the propellers 110 is one, there are few amounts of battery consumption and prolonged employment is attained as compared with a body which has a plurality of rotary wings.

[0049]

By the visual cue according to the reflected light and the streamer object 132 from a motion and the body of the body like [the small unmanned aircraft P4] the small unmanned aircraft P2 and the auditory stimulus by the whizzing sound at the time of a flight, Wildlife can be threatened visually and auditorily and a motion and intimidation means of the body are further expected by a superimposed thing [carrying out] that wildlife becomes difficult to get used to a stimulus.

[0050]

The portion from which the small unmanned aircrafts P1-P4 of the 1st to 4th embodiment make the wing tip and body produce a whizzing sound and a ****** phenomenon is formed, and it may be made to heighten the intimidation effect by a sound.

[0051]

Thus, according to the small unmanned aircrafts P1-P4 of a 1st embodiment (a modification is included) to a 3rd embodiment. It can threaten and drive off to the wildlife 51, 52, and 53 which damages a field etc. using the motion by the flight in the skies, such as a field like the farmland F, a flight sound or light, and the intimidation means of a sound.

[0052]

There is no intention which excepts a term and expression the term and expression which are used on these Claims and these Descriptions are a thing on a description to the last, restrictive expression at all and equivalent to the characteristics described by these Claims and this Description and its part. It cannot be overemphasized that various deformation modeses within the scope of technical idea of this design are possible.

[Explanations of letters or numerals]

[0053]

P1, P2, P3, and P4 Small unmanned aircraft

10 Body part

11 Rotary wing

110 Propeller

12 Leg

130 130a Reflective surface

131 Hanging object

132 Streamer object

14 Loudspeaker part

51 Crow

52 Wild boar

53 Monkey

F Farmland

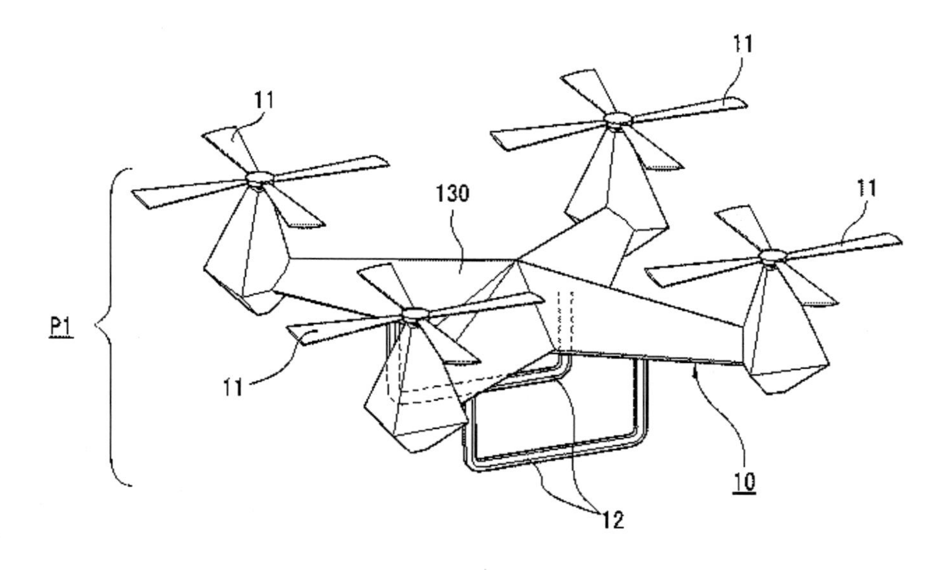

drawing 1

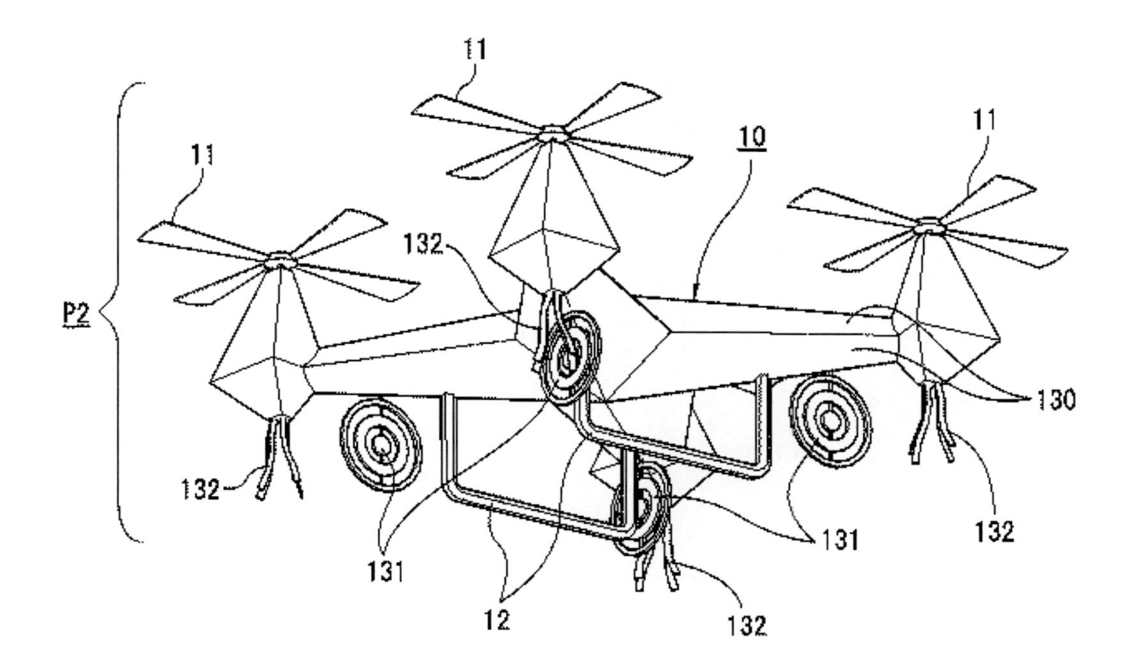

drawing 2

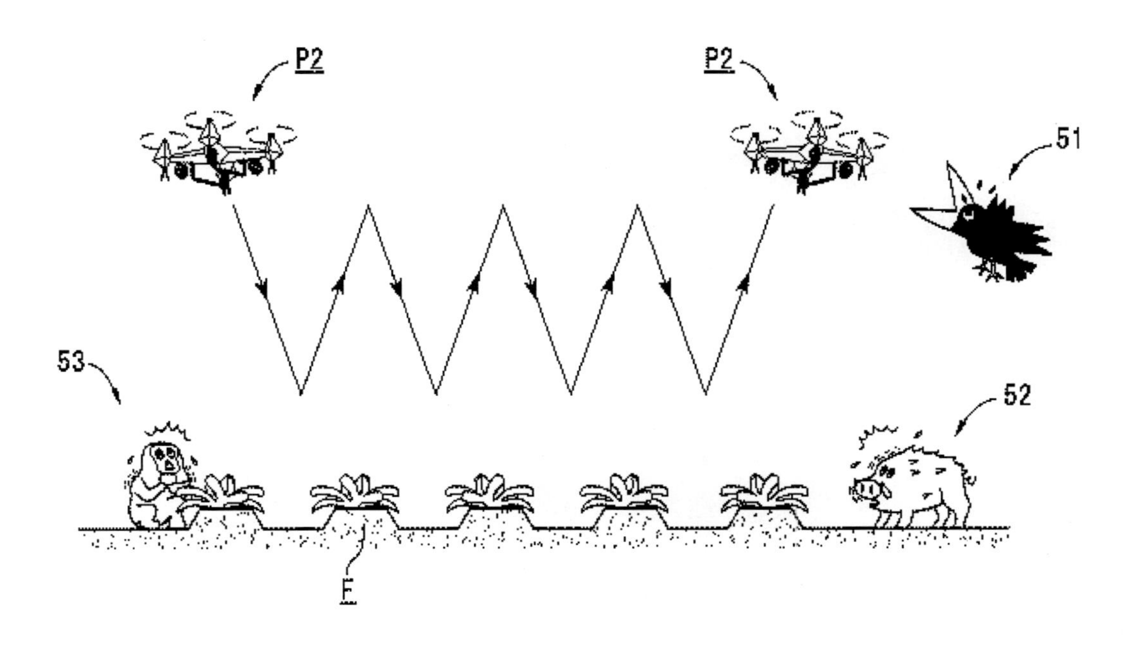

drawing 3

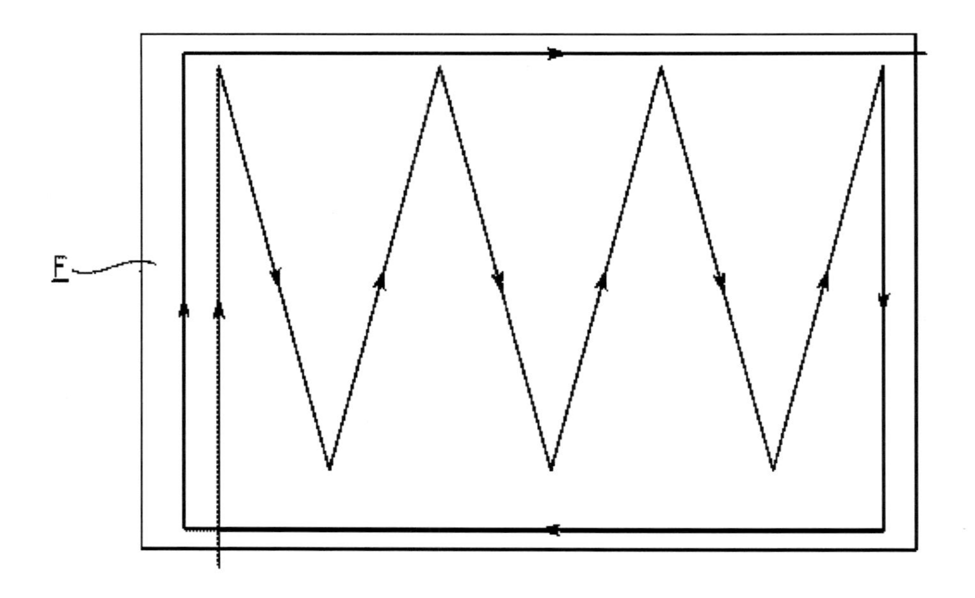

drawing 4

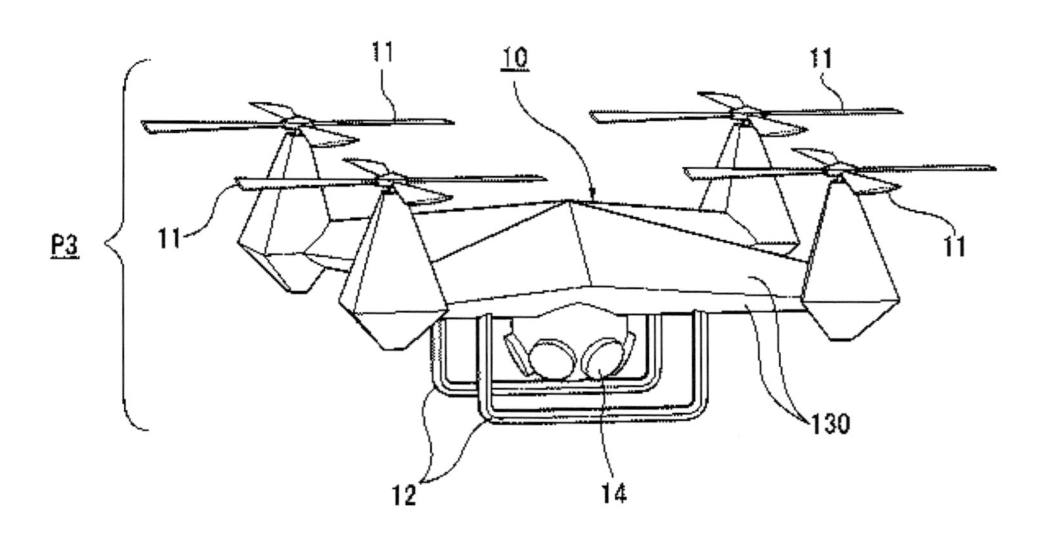

drawing 5

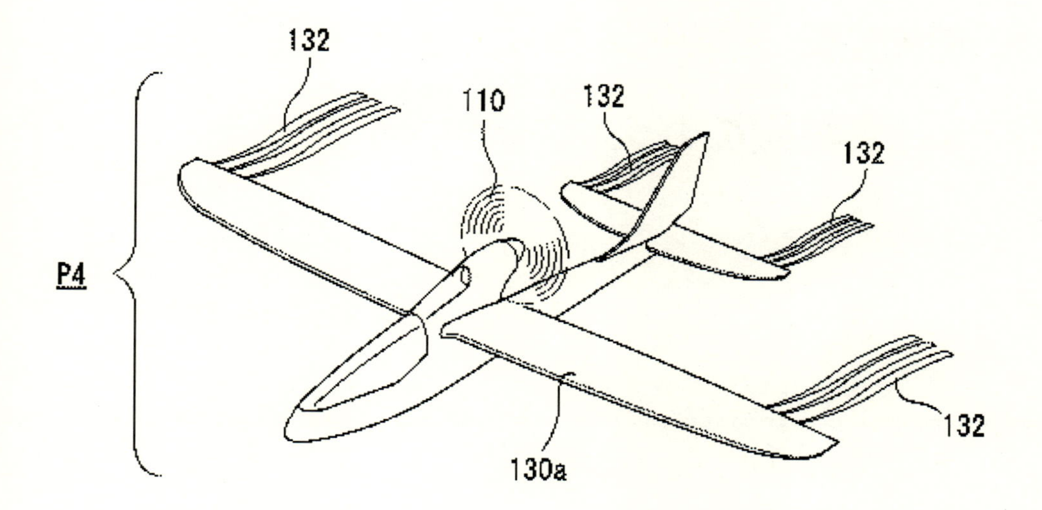

drawing 6

紫外線遮断装備 ドローン カラス撃退

スタジアム・果樹園のカラス撃退システム

定価（本体 1,500 円＋税）

２０１６年（平成２８年）８月２０日発行

No. 
HG98-1-019

発行所　IDF（INVENTION DEVLOPMENT FEDERATION）
発明開発連合会®

ﾒｰﾙ 03-3498@idf-0751.com　www.idf-0751.com
電話 03-3498-0751㈹
150-8691 渋谷郵便局私書箱第２５８号

発行人　ましば寿一
著作権企画　IDF 発明開発(連)
Printed in Japan
著者　樋口 節美©
(ひぐちせつみ)
